Generis

PUBLISHING

Beneath the Surface: Understanding Danube Pollution and Zooplankton Ecology

Mohammad-Mourad Rachad Halwani

Title: **Beneath the Surface: Understanding Danube Pollution and Zooplankton Ecology**

ISBN: 979-8-89248-550-0

Author: Mohammad-Mourad Rachad Halwani

Cover image: www.pixabay.com

Publisher: Generis Publishing
Online orders: www.generis-publishing.com
Contact email: info@generis-publishing.com

Introduction of the book

In this book, I introduce two research studies:

1. Danube River Water and Sediment Analysis

The first study involves analyzing the water and sediment from the Danube River. This research examines various physical, chemical, and biological aspects to identify potential issues not visible to the naked eye. The results have been sent to Bulgarian biosphere management authorities to address these hidden problems through continuous monitoring and timely intervention.

2. Zooplankton Community Comparison

The second study compares zooplankton communities in Lake Srebarna and Malak Preslavets Swamp. Despite different water sources for these two ecosystems, the research explores how their varying environmental conditions impact zooplankton diversity and composition. This comparison helps in understanding how different ecological factors influence aquatic life.

These studies highlight the importance of detailed environmental analysis and the value of comparative research in managing and understanding aquatic ecosystems.

TABLE OF CONTENTS

Research of the quality of water and sediments in the Silistra region of the Danube

Mohammad- Mourad Rachad Halwani
Silistra, Bulgaria
mourad@abv.bg

Inroduction:

In the ever-evolving landscape of industrial and domestic activities, their increasing impact on the fragile balance of our ecological systems cannot be ignored. This is particularly evident in the case of the Danube River, a vast and critical waterway that winds through an astounding 19 countries, standing as one of Europe's most extensive river systems. The Danube plays a vital role in supporting the livelihoods of millions, making it an indispensable natural resource. However, this importance also brings to light the significant threat posed by potential contamination from various organic and inorganic substances. The implications of high concentrations of such hazardous materials are profound, not only for the delicate ecosystems that thrive within and around the river but also for the human populations that depend on it for their daily needs.

As the project advanced, a critical challenge emerged: the absence of clear and updated standards and regulations for identifying and determining permissible concentrations of priority substances in the Danube River's sediment. While there are existing regulations for water quality, these standards have not been consistently maintained or updated over the years, leading to gaps in environmental protection. Consequently, in the project's developmental phase, the team had to rely on data from loamy sandy soils and existing European and American standards as substitutes to guide their analyses.

The primary goal of this project is to conduct a thorough assessment of the Danube River's condition in the region of Silistra. This is achieved through detailed analyses of both water and sediment samples collected from three distinct locations along the river. These analyses were performed within the highly regarded laboratories of the Institute of General and Inorganic Chemistry (IGIC), the Geological Institute (GI), and the Institute of Organic Chemistry, all part of the prestigious Bulgarian Academy of Sciences (BAS). The findings from these analyses are not merely academic; the project aims to raise awareness among the residents of Silistra about the severity of the contamination issues, while simultaneously driving forward reforms in the methodologies used for river monitoring.

To address the identified issues, the project proposes the creation of a sustainable and comprehensive monitoring system that covers both water and sediment across a wide range of sampling points along the Danube. This system would allow for continuous, real-time monitoring of the river's condition, ensuring that any signs of contamination or environmental distress can be promptly detected and addressed. Such a proactive approach would enable authorities and stakeholders to implement swift and effective interventions whenever pollutant levels exceed the established safety thresholds.

The project's foundation was laid through active participation in training sessions connected to the SIMONA European project, which provided valuable insights and methodologies for environmental monitoring. Further expertise was gained through collaboration with specialists from the Geological Institute, who played a key role in shaping the project's approach. With the support of the IGIC and GI, a detailed plan was developed for studying water and sediment in the Silistra area. The analytical work employed two distinct methodologies, enhancing the reliability and depth of the research findings. Ultimately, the project concluded with a careful comparison of the results obtained with those from the SIMONA project, ensuring that the insights gained were both rigorous and aligned with broader European efforts in environmental protection.

Through these efforts, the project not only contributes to a better understanding of the current state of the Danube River in Silistra but also sets the stage for long-term improvements in environmental monitoring and protection, helping to safeguard this vital waterway for future generations.

Project work:

1 Sampling collection

1.1 Sampling bottom and suspended sediment

In June 2021, I participated in training session as part of the SIMONA project, a collaborative cross-border initiative aimed at implementing significant reforms in the monitoring of water and river sediment across Europe. The project's primary objective is to enhance and standardize the methods used for assessing the health and quality of rivers, with a particular focus on sediment analysis. This endeavor is critical, as sediment often carries pollutants that can have long-term impacts on ecosystems and human health.

During this initial training session, I was actively involved in the sampling process, which provided us with hands-on experience in using various tools and techniques for collecting both bottom and floodplain sediment. The tools I utilized included a two-bladed spade, which is designed for efficient sediment collection, a hand spade for more precise sampling, a vacuum corrector to ensure accurate sample retrieval, and a special drill tailored for deeper sediment extraction. These methods allowed us to gather a wide range of sediment samples, which are crucial for the comprehensive analysis required to monitor and assess the condition of river systems effectively.

The practical experience gained from this session was invaluable, as it equipped us with the necessary skills and knowledge to contribute effectively to the ongoing efforts of the SIMONA project. The insights and techniques learned during this training are essential for ensuring that the monitoring of water and sediment is both thorough and consistent, helping to protect and preserve the delicate balance of Europe's river ecosystems. (Figures 1 and 2).

1.2 Sampling of suspended sediment

In November 2021, I had the opportunity to engage in specialized training on suspended sediment sampling, conducted by scientists from the Geological Institute. During this training, I utilized an advanced device known as a passive membrane sampler. This device was strategically affixed to the pontoon at the Druster Hotel harbor, a location chosen for its accessibility and relevance to our study area.

The passive membrane sampler is designed to capture suspended sediment from the water, a critical component in understanding the overall sediment dynamics and potential contamination in the river. The device operates on the principle of absorption, efficiently collecting sediment particles as water flows through it. It achieves this by filtering the suspended sediment through three distinct filters, each contributing to the thorough capture and retention of various particle sizes and types. This method provides a comprehensive snapshot of the suspended sediment load in the river, offering valuable data for assessing the environmental quality and potential pollutant levels in the water. The setup and operation of this device during the training session are illustrated in Figure 3.

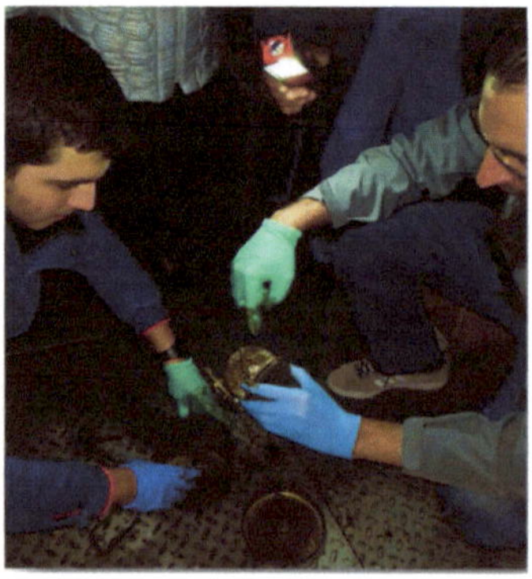

1.3 Sampling from the determined 3 points in Silistra

I undertook an independent research project to assess the water quality and sediment composition in the Silistra region of the Danube River. To determine the most relevant sampling locations, I conducted a thorough exploration of the Danube's shoreline within the city limits. During this survey, I identified a significant pollution source in the plane area, where sewage was being discharged directly into the river from an old canal. This canal historically served as the main wastewater outlet for the city. Given the evident contamination at this site, I selected it as our first sampling point.

Our second sampling location was designated in the harbor area, a site of considerable human and industrial activity, which made it a crucial point for understanding the impact of such activities on the river's health. The third sampling point was located within the Danube Park area, a location that provided a contrasting environment to the more industrially influenced sites. These three strategically chosen points offered a comprehensive overview of the varying conditions along the river in this region. The precise locations of these sampling points are illustrated on the accompanying map (Figure 7).

For water sampling, our protocol involved collecting approximately 1/2 liter of water at each site, ensuring the samples were free of residual air to prevent oxidation and other changes during storage. Additionally, I collected around 1/2 kilogram of bottom sediment from the top 5 cm layer of the riverbed at each location. The sediment was gathered using a hand spade and a grabber tool, both constructed from unpainted stainless steel. This material choice was deliberate, as stainless steel minimizes the risk of introducing any contaminants into the samples, ensuring that the data collected would accurately reflect the river's natural conditions. All samples were then stored in a refrigerated bag to maintain their integrity until analysis could be performed. The sampling process and tools used are documented in Figures 4, 5, and 6.

1.4 Sampling under the guidance of the Regional Environmental and Water Inspectorate- Ruse

In June 2022, I undertook water sampling in collaboration with the environmental agency, focusing on the Danube River at Silistra. This effort was supported by the border police, who facilitated the collection of samples from several key locations: the Bulgarian coast, the thalweg (the deepest part of the river channel), and the Romanian coast. This comprehensive sampling approach aimed to capture a representative cross-section of water quality across different parts of the river.

Once the samples were collected, they were transported to a laboratory in Ruse for detailed analysis. This analysis was crucial for evaluating the water quality and detecting any potential contaminants or pollutants present in the river. The laboratory in Ruse is equipped with the necessary facilities and expertise to perform thorough testing, ensuring that the data gathered would provide accurate and actionable insights into the state of the Danube River in the Silistra region.

2. Sample analysis

2.1 Inorganic chemical analysis in the Institute of general and inorganic chemistry

After collecting the water samples, the initial step in the analysis process involves filtering the samples to remove any particulate matter. The properties of the filtered water are then assessed using a multimeter, a versatile instrument that measures various parameters including pH, conductivity, and dissolved oxygen. These parameters are crucial for understanding the basic chemical and physical characteristics of the water, which are essential for evaluating its quality and identifying potential contamination.

For more detailed analysis of specific parameters, spectrophotometry is employed. This technique involves mixing a precise amount of the water sample with reagents that react to form a colored solution. This solution is placed in a cuvette, which is then inserted into a spectrophotometer. The spectrophotometer exposes the solution to light of a specific wavelength. The interaction between the light and the colored solution produces a spectrum that is used to determine the concentration of various nutrients and substances in the water. By measuring the absorbance of light at particular wavelengths, the spectrophotometer provides numerical readings that indicate the concentration of the target analytes.

The analysis of sediment for heavy metals involves several meticulous steps. Initially, sediment samples are air-dried and then subjected to lyophilization, a process that freezes the water within the sediment and converts it directly into gas through sublimation, leaving behind a dry, powdered sample. The lyophilized sediments are then ground to a fine powder and sieved through a 1 mm sieve to ensure uniformity in particle size.

To determine the heavy metal content, a portion of the lyophilized sediment, typically 0.5 grams, is subjected to digestion using a mixture of concentrated nitric acid and hydrochloric acid (aqua regia) in a 1:3 ratio. This digestion process, conducted under reflux conditions for three hours, breaks down the sediment and releases the metals into solution. After digestion, the solution is filtered, and the resulting liquid is quantitatively transferred and diluted with water to a final volume of 100 mL, in accordance with the ISO 11466 standard.

The prepared samples are then analyzed using the Prodigy7 instrument, which employs Inductively Coupled Plasma Optical Emission Spectrometry (ICP-OES). ICP-OES is a widely used analytical technique for determining the atomic composition of a sample. The technique involves exciting the sample in a plasma state using an inert gas, typically argon. As the sample's molecules break down into atoms, they emit electromagnetic radiation at characteristic wavelengths. By analyzing these emitted wavelengths, the ICP-OES system identifies the specific elements present in the sample and quantifies their concentrations based on the intensity of the emitted light. This comprehensive approach enables precise detection and quantification of various heavy metals and other elements within the sediment, providing critical information about the contamination levels and overall environmental health of the area being studied.

2.2 Geochemical analysis in the Geological institute of Bulgaria

A sediment sample from the first point was analyzed.

To ensure that the control sample accurately represents the entire sediment sample, a rigorous preparation process is followed. Initially, the sediment is air-dried to remove any moisture. It is then homogenized to achieve a uniform consistency, and divided into quarters to ensure that each portion is representative of the whole sample. These quarters are subsequently ground into a fine powder to facilitate consistent and precise analysis.

For analyzing the mineral composition of the control sample, X-ray diffraction (XRD) is employed. The analysis is conducted using a PANalytical Empyrean XRD apparatus, which is equipped with a multichannel detector (Pixel 3D). The XRD process utilizes Cu Kα radiation (45 kV, 40 mA) to probe the sample. The diffraction pattern is recorded over a 2θ range of 4° to 90°, with a scan step size of 0.0394° and a counting time of 45.135 seconds per step. This detailed scan allows for the identification and quantification of various minerals present in the sediment.

To determine the elemental content of the control sample, a different analytical technique is employed. The sediment powder is first transformed into a molten pellet by mixing it with lithium tetraborate. This mixture is subjected to high temperatures, approximately 1000°C, within a specialized crucible. The high-temperature melting process ensures that the sample is thoroughly homogenized and allows for accurate determination of its elemental composition.

Once the sediment sample has been transformed into a molten pellet, it is analyzed using a 193 nm ArF excimer laser (ATLEX-LR, Germany), which is coupled with an ELAN DRC-e Inductively Coupled Plasma Mass Spectrometry (ICP-MS) instrument (Perkin Elmer, Canada). The excimer laser is used to ablate the molten pellet, and the resulting material is introduced into the ICP-MS for detailed analysis. The ICP-MS technique measures the elemental composition by detecting ions generated from the ablated sample. This method provides precise quantification of the various elements present in the sediment, offering comprehensive insights into its chemical composition and contamination levels.

2.3 Geoaccumulation Index

The Geoaccumulation Index was employed to assess the extent of heavy metal contamination in the sediments analyzed at IGIC. The index provides a quantitative measure of the contamination levels and allows for a comparative analysis of different sampling points.

The geoaccumulation index is denoted I_{geo} as:

$$I_{geo} = \log_2\left(\frac{C_n}{1.5 \times B_n}\right)$$

Where:

concentration of the element in the sediment- C_n

the concentration of the element in the earth's crust- B_n

Depending on the values of I_{geo}, the sediments are divided into groups

zone A (<0) – uncontaminated;

zone B (0 - 1) – unpolluted or moderately polluted;

zone C (1 - 2) - moderately polluted;

zone D (2 - 3) - moderately to heavily polluted;

zone E (3 - 4) - heavily polluted.

2.4 Organic chemistry analysis in the Organic chemistry institute

Sediment samples from the first and second point were analyzed in order to determine the organic contaminants. The samples were analyzed with 2 methods: Gas Chromatography-Flame Ionization Detection (GC-FID) and Gas Chromatography-Mass Spectrometry (GC- MS).

n GC-FID, a sample is first injected into the gas chromatograph, where it is vaporized and carried by an inert gas, usually helium or nitrogen, through a chromatographic column. The column contains a stationary phase that interacts with the different components of the sample. As the sample passes through the column, these components separate based on their volatility and interactions with the stationary phase. Each compound exits the column at a different time, known as the retention time, allowing for their individual identification and quantification.

Once the separated compounds exit the column, they enter the flame ionization detector (FID). In the FID, the compounds are burned in a hydrogen-air flame, causing the organic molecules to ionize. These ionized molecules generate an electrical current that is proportional to the number of carbon atoms in the compound. The intensity of this current is measured and recorded, producing a signal for each compound. By analyzing these signals, the concentrations of the compounds in the sample can be determined.

In the GC-MS process, a sample is first introduced into the gas chromatograph, where it is vaporized and mixed with a carrier gas, typically helium or nitrogen. The vaporized sample is then carried through a long, narrow column that contains a stationary phase material. As the sample travels through the column, its different components separate based on their volatility and their interactions with the stationary phase. The components that are more volatile move faster, while those that interact more strongly with the stationary phase move more slowly. This separation allows the individual compounds within the mixture to exit the column at different times, creating a series of peaks that can be detected and analyzed.

Once the compounds exit the gas chromatography column, they enter the mass spectrometer. In the mass spectrometer, the compounds are ionized, typically by an electron impact, which causes them to break into charged fragments. These fragments are then sorted based on their mass-to-charge ratio (m/z) in the mass analyzer. The mass spectrometer produces a mass spectrum for each compound, which is essentially a unique "fingerprint" that can be used to identify the compound. By comparing the mass spectrum to a library of known spectra, the specific compounds present in the sample can be identified.

3. Results

3.1 Results from the water samples analysis

1. Analysis of the data reveals that the highest values for almost all indicators listed in Table 1 were observed at sampling point 1. However, it is noteworthy that only phosphates in samples 1 and 3 exceeded the maximum permissible concentrations. In sample 1, specifically, the phosphate levels were found to be 3.7 times higher than the maximum permissible concentration.
2. The water samples demonstrate overall cleanliness in terms of the tested metals. The only exception is the manganese concentration, which surpassed the maximum permissible limit at points 1 and 2, as indicated in Table 2.
3. Notably, the water quality is found to be the best near the Danube Park, while the highest level of contamination is observed at the sampling point adjacent to the canal. This information, derived from the results, highlights the spatial variation in water quality along the studied section of the Danube River.
4. The water analysis of samples collected from the Danube River has revealed significant insights into the presence and concentrations of heavy metals. These findings, presented in Table 2, highlight the measured values of specific heavy metals and their implications for the ecosystem and human health. Heavy metals, such as lead (Pb), cadmium (Cd), arsenic (As), among others, are naturally occurring elements with high atomic weights that can be harmful to living organisms when found in elevated levels. They can enter water bodies through various sources, including industrial discharges and agricultural runoff.

Table1 Note:(a) Bulgarian regulation

Sample number	Maximum allowable concentrations	Point 1- The channel	Point 2- The port	Point 3- The Danube Park
T°C		15	15	15
pH	6,5 – 8,5 [a]	8,43	8,47	9,31
Eh, mV				
Conductivity, μS/cm	750 [a]	560	443	435
Dissolved O_2,	7.00-6.00 [a]	6,9mg/l 66,1%	6,5mg/l 67,1%	8,63mg/l 84,4%
Salinity		<0,01	<0,01	<0,01
Alkalinity, mg/l		201	132	145
Main elements				
Cl, mg/l	300	30	22	22
SO_4^{2-}, mg/l	300	17,4	28,5	21,2
CO_3^{2-}, mg/l		120	79,2	87
Ca, mg/l		69	52	58
Nutrients				
PO_4^{3-}, mg/l	0.07-0.15 [a]	0,55	0,06	0,18
$N\text{-}NO_2^-$, mg/l	0.03-0.06 [a]	0,03	0,01	0,01
$N\text{-}NO_3^-$, mg/l	0.7 - 2 [a]	1,6	1,3	1,0
$N\text{-}NH_4^+$, mg/l	0.1-0.3 [a]	0,10	0,08	0,06

4/2012- "water II quality" for plain types of rivers

Table 2

Sample number	Maximum allowable concentrations	Point 1- The channel	Point 2- The port	Point 3- The Danube Park
Content in water samples, mg/l				
Fe	0.1 [a]	0.046	0.022	0.033
Mn	0.05 [a]	**0.099**	**0.075**	0.040
Cu	0.01 [a]	0.009	0.006	<0.005
Zn	0.075 [a]	0.010	0.014	<0.005
Cd	0.0015 [b]	<0.005	<0.005	<0.005
Pb	0.014 [b]	<0.05	<0.05	<0.05

Note:a REGULATION No. H-4 of 14.09.2012 on characterization of surface waters

b DIRECTIVE 2013/39/EU OF THE EUROPEAN PARLIAMENT AND OF THE COUNCIL of 12 August 2013

3.2 Results from the sediment samples analysis

1. The results of the sediment analysis are presented in Table 5. In the column for the allowable concentrations for sediment, values for clayey-sandy soils are provided, as there are no specific Bulgarian standards or regulations for Danube sediment concentrations and clayey-sandy soils have a composition that is closest to sediments.
2. Based on the data presented in Table 5, it can be concluded that the concentration of the cadmium metal in the first point exceed the allowable concetration almost five times.

3. Among the various metals analyzed, iron (Fe) and manganese (Mn) were found to be present in the highest amounts in the sediments. These elements are naturally occurring in different silicate substances that constitute the sediments. Therefore, they cannot be categorized as pollutants in the sediment samples. The higher concentrations of Fe and Mn can be attributed to their inherent composition and prevalence in the sediments.
4. However, it is crucial to note that the concentration of cadmium exceeds the permissible limit. This finding raises concerns as it indicates potential contamination of the sediments with cadmium. The presence of elevated cadmium levels should be taken into account and addressed to mitigate any associated environmental risks.
5. The presence of oxide compounds in the sediment is closely related to the minerals that are present. The predominant oxides observed are SiO2 and Al2O3. Increased concentrations of CaO and MgO are also detected, indicating their binding to carbonates. The oxides of alkali elements are primarily associated with minerals such as albite and muscovite, as indicated in *(table 6)*.
6. The trace elements measured in the sediment fall within the expected range for river bottom sediments. Notably, heavy metals including vanadium (V), chromium (Cr), cobalt (Co), nickel (Ni), zinc (Zn), cadmium (Cd), and lead (Pb) are present in significant amounts. Additionally, barium (Ba) and strontium (Sr), which are typically associated with carbonate minerals, are also detected. All rare earth elements (REEs) are present in the sample as well. However, the content of 15 elements is below the detection limit of the analysis as shown in *(tables 7 and 8)*.
7. By utilizing X-ray diffraction (XRD), the primary minerals present in the sediment sample are identified. The analysis reveals that the predominant mineral components are reflected in the content of major element oxides, with silicon dioxide (SiO_2) and aluminum oxide (Al_2O_3) being the most prevalent. The minerals identified include quartz, various carbonates such as dolomite and calcite, as well as apatite and birnessite. These findings are illustrated in Diagram 2, which provides a visual representation of the mineral composition and the presence of key elements in the sediment sample.

*Table 5 *Not detected*

Sample number	Allowable concentrations	Point 1- The channel	Point 2- The port	Point 3- The park
Content in sediment samples, mg/kg				
Fe		28206	30034	26727
Mn		886	740	621
Cu	50	54.8	88.1	54.0
Zn	110	145	168	103
Cd	0.6	3.11	1.67	*
Pb	40	42.7	66.2	24.1

Table 6

SiO_2	wt.%	58.65
TiO_2	wt.%	0.65
Al_2O_3	wt.%	12.62
FeO	wt.%	3.66
MnO	wt.%	0.12
MgO	wt.%	2.45
CaO	wt.%	6.71
Na_2O	wt.%	1.60
K_2O	wt.%	1.84
P_2O_5	wt.%	0.17

Table 7

Be9	µg/g	<51.2355
Sc45	µg/g	13.93
V51	µg/g	**99.69**
Cr53	µg/g	108.23
Co59	µg/g	14.30
Ni60	µg/g	56.45
Cu65	µg/g	69.29
Zn66	µg/g	140.15
Ga71	µg/g	14.98
Ge74	µg/g	<8.3507
As75	µg/g	15.24
Se77	µg/g	<34.9616
Se82	µg/g	<131.0002
Rb85	µg/g	81.53
Sr88	µg/g	181.87
Y89	µg/g	23.71
Zr90	µg/g	218.47
Nb93	µg/g	12.07
Mo95	µg/g	<8.0173
Pd105	µg/g	<2.2625
Ag107	µg/g	<2.9675
Cd111	µg/g	26.52
In115	µg/g	<0.3197
Sn118	µg/g	<5.557
Sb121	µg/g	<2.5244
Te125	µg/g	6.08
Cs133	µg/g	5.16

Table 8

Ba137	µg/g	419.18
La139	µg/g	28.91
Ce140	µg/g	59.04
Pr141	µg/g	6.11
Nd146	µg/g	25.44
Sm147	µg/g	4.89
Eu151	µg/g	1.10
Gd157	µg/g	5.49
Tb159	µg/g	0.82
Dy163	µg/g	4.85
Ho165	µg/g	0.83
Er167	µg/g	2.64
Tm169	µg/g	0.51
Yb173	µg/g	<1.902
Lu175	µg/g	<0.4207
Hf178	µg/g	5.37
Ta181	µg/g	1.07
W182	µg/g	2.31
Re185	µg/g	<1.1219
Pt195	µg/g	37.13
Au197	µg/g	<1.1677
Tl205	µg/g	<0.97218
Pb208	µg/g	37.81
Bi209	µg/g	0.65
Th232	µg/g	8.76
U238	µg/g	2.67

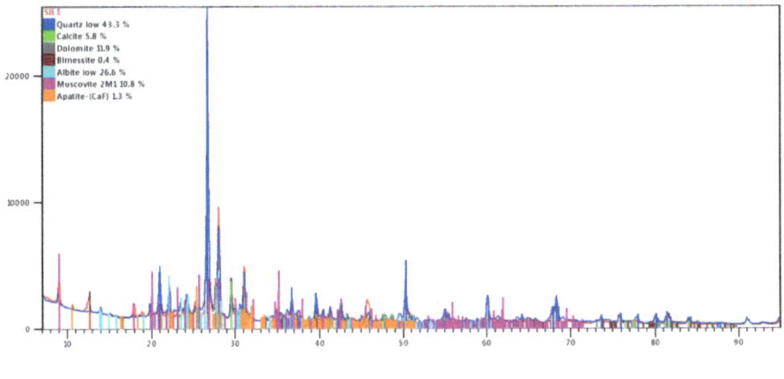

Diagram 2

3.3 Conclusions form Igeo (Diagram 1)

1. Fe, Mn, and Cu in points 1 and 2, as well as Zn in point 3, fall within zone A, indicating no contamination with these metals. Their levels are within acceptable limits.
2. Zn in points 1 and 2 is in zone B, suggesting either no or moderate contamination. Although present, Zn remains within acceptable levels.
3. Cd in point 1 is heavily polluted (zone E), while in point 2 it is moderately to heavily polluted (zone D). These higher concentrations of Cd indicate environmental risks, requiring further investigation and remediation.
4. Pb at point 1 is moderately contaminated (zone C), at point 2 it is moderately to heavily polluted (zone D), and at point 3 it is either uncontaminated or slightly contaminated (zone B). The contamination levels of Pb vary across the sampling points.

In summary, the Igeo diagram provides valuable information on metal contamination in the sediment. Fe, Mn, Cu, and Zn are within acceptable limits, while Cd and Pb show varying degrees of pollution. These findings inform decision-making and the implementation of appropriate measures to manage and mitigate metal contamination in the studied areas.

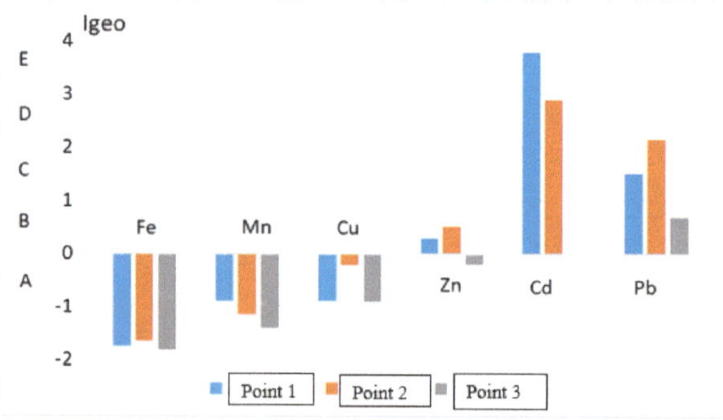

Diagram 1

3.4 Conclusions from the organic analysis (table 9)

1. Phthalates have high concentration. They are known to be toxic and can act as endocrine disruptors in aquatic environments. The high concentrations of phthalates in both waste water samples are concerning.

2. Triacetin has elevated concentration. While generally considered to have low toxicity, the very high concentrations suggest that it could impact the biochemical oxygen demand (BOD) and chemical oxygen demand (COD) in the water, potentially leading to depletion of dissolved oxygen necessary for aquatic life.
3. By and large phthalates are particularly concerning due to their high concentrations and known toxicological effects on aquatic life. Triacetin, despite its lower toxicity, is present at very high concentrations, which could also pose environmental risks due to its impact on oxygen levels in the water. Both of these compounds warrant closer monitoring and potential remediation to protect river water quality.

Table 9

No	Compounds	Formula	Mw	Rel.%, as determined by GC-FID	
				WASTE WATER-1	WASTE WATER-2
1.	Butanoic acid 3-methyl 3-methylbutyl ester (=Isopentyl isovalerate)	$C_4H_8O_2$	88	8.43	14.28
2.	1-Dodecene	$C_{12}H_{24}$	168	1.51	3.49
3.	Glycerol, 1,2-diacetate	$C_7H_{12}O$	176	2.10	2.04
4.	1,2-Benzothiazole			0.31	0.39
5.	Triacetin (= 1,2,3 Propanetriol triacetate)	$C_9H_{14}O_6$	218	39.83	41.44
6.	Eugenol	$C_{10}H_{12}O_2$	164	1.38	2.09
7.	Tetradecene	$C_{14}H_{28}$	196	1.21	2.56
8.	Methyl cinnamate or 4-Vinylbenzohydrazide)	$C_{10}H_{10}O_2$	162	2.81	4.62
9.	2-Furanone,5-hexadihydro(=Peach lactone = γ-undecalactone)	$C_4H_4O_2$	84	0.34	0.50
10	Phenol,2,4-olitertbutyl	$C_{14}H_{22}O$	206	0.26	0.51
11.	Tetradecene (isomer)	$C_{14}H_{28}$	196	0.32	0.58
12.	Buthyl benzoate	$C_{11}H_{14}O_2$	178	0.43	0.54
13.	Benzoic acid, decyl ester	$C_7H_6O_2$	122	0.52	0.63
14.	Phthalate	$C_8H_4O_4$	164	0.63	0.79
15.	Benzoic acid, unknown ester			1.28	1.92
16.	Benzoic acid, unknown ester			1.60	1.86
17.	Benzoic acid, unknown ester			1.13	1.30
18.	Phthalate	$C_8H_4O_4$	164	1.15	0.86
19.	Tributhyl aconitate (=1,2,3-Propentricarbxylic acid tributil ester)	$C_{18}H_{30}O_6$	342	1.24	1.63
20.	Phthalate			6.31	9.07

4. Comparison of the results

4.1. Comparison with the results from the National Monitoring 2017

I compared the water analysis results obtained from the Institute of General and Inorganic Chemistry (IGIC) with those from the national monitoring conducted in 2017, as presented in Table 3. The samples for both analyses were collected from the same points in Silistra, ensuring consistency in the locations studied.

Upon examining the data from Table 2 and Table 3, it can be inferred that the concentration of metals in the water samples has remained relatively stable over the monitored period. This stability suggests that there have been no significant fluctuations in metal levels between the two sets of results. The consistency in metal concentrations over several years indicates that, during this time frame, the levels of these pollutants have not experienced considerable changes, reflecting a steady state of metal contamination in the river water at the specified locations.

Table 3

Sample number	Date of sampling	Fe	Cd	Mn	Cu	Pb	Zn
		mg/l	mg/l	mg/l	mg/l	mg/l	mg/l
Point 1- The channel	4/11/2017	<0.015	<0.0005	0.0102	0.0032	<0.0004	<0.005
Point 2- The port	4/11/2017	<0.015	<0.0005	0.0014	0.0018	<0.0004	0.0157
Point 3- The Danube park	4/11/2017	<0.015	<0.0005	0.0010	0.0018	<0.0004	<0.005
Maximum allowable concentrations		0.1a	0.0015 b	0.05 a	0.01 a	0.014 b	0.075 a

4.2. Comparison with the results from the Regional Environmental and Water Inspectorate- Ruse (table 4)

An examination of both Table 2 and Table 4 reveals that no substantial differences in the concentrations of the investigated elements are observed. This finding indicates a consistent pattern across the various sources and analytical methods employed.

Table 4

	Bulgarian cost	Thalweg	Romanian coast
Fe	0.008 mg/L	0.008 mg/L	0.008 mg/L
Zn	0.003 mg/L	0.003 mg/L	0.007 mg/L
Cu	0.002 mg/l	0.002 mg/l	0.002 mg/l
Cd	<0.005 mg/L	<0.005 mg/L	<0.005 mg/L
Pb	<0.05 mg/L	<0.05 mg/L	<0.05 mg/L
Mn	0.002 mg/L	0.005 mg/L	0.005 mg/L

4.3. Comparison with the obtained results from SIMONA project for Silistra

To assess the level of sediment contamination, conducted a comparison between the concentrations of heavy metals in the bottom sediment of the analyzed channel at the Geological Institute and the results obtained from the SIMONA study conducted in Silistra. This comparison was made with reference to European and American standards, as depicted in charts 1, 2, 3, 4, 5, and 6.

The charts presented illustrate the comparison between the sediment samples obtained from the channel and the findings of the SIMONA project. It is noteworthy that elevated concentrations of all elements were observed when comparing the results with those obtained from the Silistra area in the SIMONA study, as well as with certain standards. As a conclusion, the metals analyzed in the channel exhibited higher values compared to the European standards.

1. The concentration of Cd is elevated relative to the American standard;
2. The concentration of Cr, Zn and Pb is increased compared to the European standards.
3. Ni concentration is elevated compared to European standards;
4. In conclusion, there are elevated values of heavy metal concentrations in the canal sample.

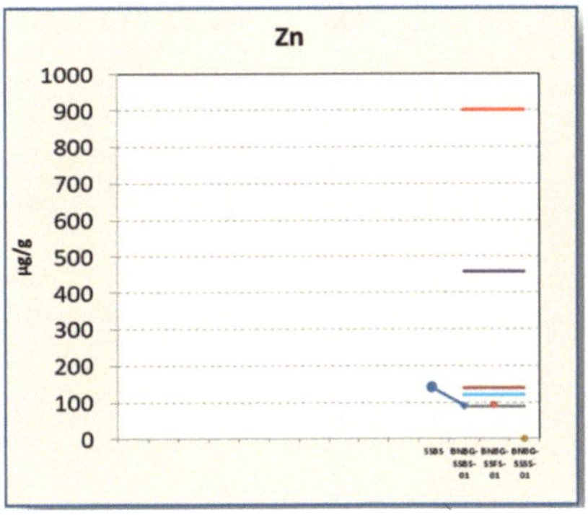

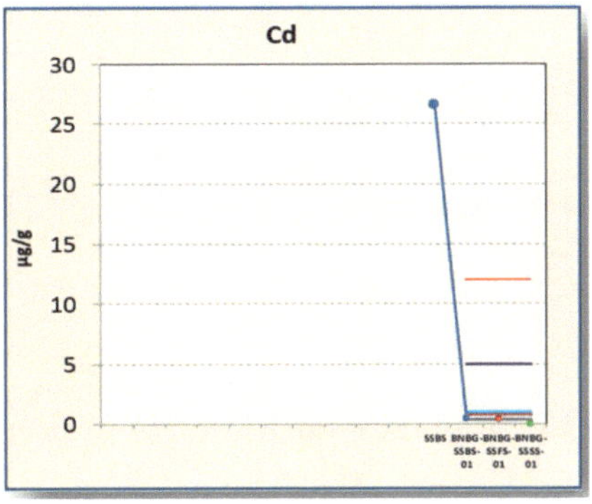

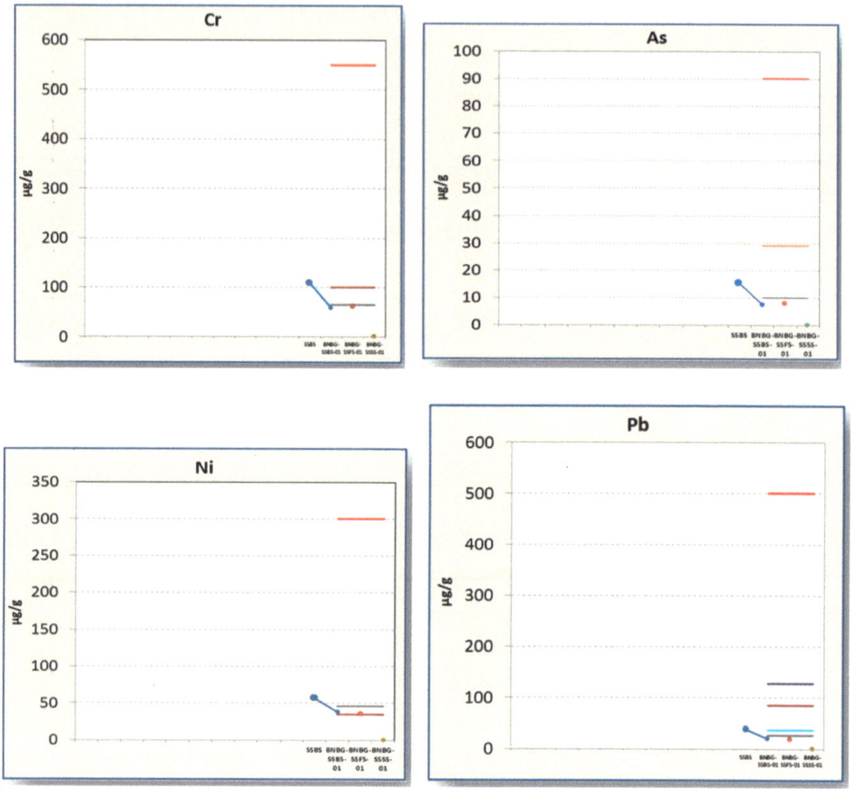

Note: - flooded sediment from SIMONA

- suspended sediment from SIMONA

- bottom sediment from SIMONA

- a sample of the canal examined in The Geological Institute BAS

- European standards

- American standards

4.4 Conclusions on the SIMONA project

Our conclusions match those obtained from the SIMONA project, which are:

1. Based on bottom, suspended, and floodplain sediment samples, there is a moderate risk to benthic or aquatic ecosystems from these metals at the time of sampling;
2. There is no obvious industrial source of metals in the nearby area of the Danube River near Silistra;
3. Thus, other relevant emission sources should be monitored from time to time, such as industrial or domestic wastewater treatment plants in the catchment. It is noted

that the main source of cadmium is fertilizers used in the EU, but not industrial activities.

5. Discussion

The results obtained from this study are of immense value to environmental scientists, policymakers, and stakeholders engaged in water resource management. They provide essential insights into areas experiencing pollution and are critical for developing effective pollution control strategies. By identifying and quantifying heavy metal contamination, these findings help in assessing the potential risks posed to ecosystems and human health. Such information is crucial for formulating targeted interventions aimed at mitigating the harmful effects of pollution.

Addressing the problem of heavy metal pollution effectively requires a well-structured approach. One proposed solution is the implementation of a comprehensive and continuous monitoring system that encompasses both water and sediment. This monitoring system would play a pivotal role in establishing clear standards and regulations for hazardous substances, including precise thresholds for their concentrations. By maintaining vigilant oversight through continuous monitoring, it becomes possible to detect any increases in pollution levels promptly and to take appropriate actions to address them.

A robust monitoring system would not only facilitate real-time assessment of water quality but also enable proactive management strategies. Timely interventions based on accurate and up-to-date data would help in preventing further environmental degradation and ensuring the long-term preservation of water resources. Such a system would be instrumental in maintaining the health of aquatic ecosystems and safeguarding public health, ultimately supporting sustainable water resource management in the region.

6. Sources

1. Avigliano, E., Monferran, MV, Sanchez, S., Wunderlin, DA, Gastaminza, J., Volpedo, AV, 2019.Distribution and bioaccumulation of 12 trace elements in water, sediment and tissues of the main fishery from different environments of the La Plata basin (South America): risk assessment for human consumption. Chemosphere 236, 124394;

2. Baltas, H., Kiris, E., Sirin, M., 2017.Determination of radioactivity levels and heavy metals;

3. Burada, A., T¸ opa, CM, Georgescu, LP, Teodorof, L., Nastase, C., Seceleanu-Odor, D., et al., 2014.Heavy metals accumulation in and water of four aquatic complexes from the Danube Delta area. Aquacult. Aquarium Conserv. Legis. 7, 301e310;

4. Danubius, RI, 2016. International Center for Advanced Studies on River-Sea Systems.

5.Brochure Danubius V4.

6.Frankowski M, Anetta ZF, Kowalski A, Siepak J (2010) Fractionation of heavy metals in bottom sediments using Tessier procedure. Environ Earth Sci 60: 1165-1178.

7.Ackerman F (1980) A procedure for correcting the grain size effect in heavy metal analyses of estuarine and coastal sediments. Environ.

8.Technol Lett 1: 518-527.

9.Allen JRL, Rae JE (1987) Late Flandrian shoreline oscillations in the Severn Estuary: a geomorphological and stratigraphical reconnaissance.

10.Philos Trans R Soc B Biol Sci 315: 185-230.

11.Loring DH (1990) Lithium - a new approach for the granulometrical normalization of trace metal data. Mar Chem 29: 156-168.

12.Loring DH, Naes K, Dahle S, Matishow GG, Illin G(1995) Arsenic, trace metals, and organic micro contaminants in sediments from the Pechora Sea, Russia. Mar Geol 128: 153-167.

13.Balls PW, Hull S, Miller BS, Pirie JM, Proctor W (1997) Trace metal in Scottish estuarine and coastal sediments. Mar Pollut Bull 34: 42-50.

14.Ryan JD, Windom HL (1988) A geochemical and statistical approach for assessing metal pollution in coastal sediments. Metals in Coastal 15.Environments of Latin America. Springer Berlin Heidelberg 47-58.

15. Sinex SA, Wright DA (1988) Distribution of trace metals in the sediments and biota of Chesapake Bay. Mar Pollut Bull 19: 425-431.

16.Emmerson RHC, O'Reilly-Wiese SB, Macleod CL, Lester JN (1997) A multivariate assessment of metal distribution in intertidal sediments of the Blackwater Estuary, UK. Mar Pollut Bull 34: 960-968.

17.Lee CL, Fang MD, Hsieh MT (1998) Characterization and distribution of metals in surficial sediments in Southwestern Taiwan. Mar Pollut Bull 36:464-471.

18.Barbieri M, Sappa G, Vitale S, Parisse B, Battistel M (2014) Soil control of trace metals concentrations in landfill: A case study of the largest landfill in Europe, Malagrotta, Rome. J Geochem Explor 143: 146-154.

19.Martin JM, Whitfield M (1983) The significance of the river input of chemical elements to the ocean. In: Wong CS, Boyle E, Brul KW, Burton JD, Goldberg ED Trace Metals in Sea Water. NATO Conference Series 9:265-296.

20.Taylor SR, McLennan SM (1995) The geochemical evolution of the continental crust. Rev Geophys 33: 241-265.

Comparative characteristics of the zooplankton communities in Srebarna lake and Malak Preslavets swamp

Introduction

The current project builds upon the foundational work conducted in a previous study focused on the examination of zooplankton within the Malak Preslavets swamp. This new phase of research aims to expand our understanding by providing a comparative characterization of the zooplankton communities present in both Lake Srebarna and the Malak Preslavets swamp during 2019, 2020 and 2021. The objective is to utilize scientific methods to analyze and compare the two ecosystems.

Methods and Materials Employed

In executing this project, we employed a comprehensive set of methods and materials to ensure thorough and accurate results. Initially, we engaged with existing scientific literature to establish a solid foundation of knowledge. Subsequently, we undertook the process of sampling from the designated locations, followed by meticulous analysis of the collected samples. To present our findings in a clear and structured manner, we created detailed tables and diagrams. Throughout this process, a scientific approach was rigorously applied in the interpretation of the data, ensuring that the results were both reliable and insightful.

Observations and Conclusions

Upon examining the data, it was observed that the majority of species at both sites belonged predominantly to the Rotatoria group. The comparative analysis of species variety diagrams for Srebarna Lake and Malak Preslavets Swamp revealed striking similarities, despite the two ecosystems being sustained by different water sources. By applying Sorensen's method and utilizing his coefficient, we discovered an increase in species similarity over time.

For the purpose of this study, we began with data from the year 2019. However, in order to extend our research, it was essential to obtain permission from the Ministry of Environment and Water to access Lake Srebarna by boat, given its status as a reserve. Our fieldwork was conducted over two visits to these protected areas between May and August, during which we collected samples for subsequent analysis. These samples were subjected to microscopic examination, and species were counted using a specialized device known as a "camera," all under the expert guidance of our scientific consultant, Veselka Tsavkova, an assistant researcher at the Institute of Biodiversity and Ecosystem Research (IBER).

Additionally, we constructed population and biomass charts, which further enriched our analysis. We also compiled summaries that explored the relationship between species diversity and the varying means of water supply in the two protected areas. In gaining familiarity with Sorensen's method for determining species similarity via a specific coefficient, we were able to apply the data from the current year alongside data from 2006, provided by our supervisor. This allowed us to conduct a comparative analysis, revealing how species similarity has evolved over a span of 15 years.

Tasks and Stages of the Project:

Study of Hydrological Features:

The first task involves a thorough investigation of the hydrological characteristics of Lake Srebarna and the Malak Preslavets swamp. This stage focuses on identifying and comparing the similarities and differences in the water sources, flow patterns, and overall aquatic environment of these two protected areas.

Sample Collection:

The next stage involves the systematic collection of water samples from both Lake Srebarna and the Malak Preslavets swamp. This task is critical for obtaining the biological material necessary for subsequent analysis.

Microscopic Observations and Hydrobiological Analysis:

Once the samples are collected, they undergo microscopic examination to identify and analyze the zooplankton species present. This includes working with a determinant to accurately identify species and using a "camera" (a specialized vessel) to count the zooplankton.

Development of Diagrams:

In this stage, the data gathered from the microscopic observations is used to create detailed diagrams that illustrate species diversity, abundance, and biomass within the two ecosystems. These visual representations help to summarize and communicate the findings effectively.

Introduction to the Sorensen Method:

An important part of the project is becoming acquainted with the Sorensen method, which is used to calculate species similarity between the two locations. This method allows for a quantitative comparison of the zooplankton communities in Lake Srebarna and the Malak Preslavets swamp.

Conclusions:

The final stage involves drawing conclusions based on the data and analyses conducted throughout the project. This includes interpreting the results, understanding the ecological implications, and identifying any trends or patterns observed in the species similarity and hydrological features of the two protected areas.

Project work

1. Hydrological features of Malak Preslavets swamp Protected area

The Malak Preslavets swamp(fig.1 and 2) is the second largest protected area in the Silistra region. It is located 30 km away from Lake Srebarna near the Danube River. There is the largest population of White Water- Lilies in Bulgaria and is rightly called the kingdom of water lilies. To get acquainted with the hydrological features of the swamp we visited the protected area. We made observations of the dike, built in 1997 (fig.3). It breaks the natural relation of the swamp with the Danube River – i.e. the relation is one-way and water from the lake, which is at higher altitudes flows into the river. This could lead to its entanglement, but the opposite happens, as the surface and underground inflows are significant. At the southwestern end of the reservoir springs the water of a river that originates from micro-dams located in neighboring villages. Underground karst springs are also active.

Beyond its ecological significance, the protected area also holds considerable historical importance due to the presence of a complex of archaeological sites. These include prehistoric settlements from the Early Neolithic and Bronze Age, a Thracian settlement, a Roman and early medieval Bulgarian fortress, and an old Roman road. This combination of ecological and historical features underscores the dual significance of the Malak Preslavets Protected Area.

In a biodiversity study conducted in 2004, the following observations were made:

Birds: The most abundant group of animals in the protected area is the birds, with a total of 147 species recorded. Of these, 82 species are confirmed as nesting within the area. Notably, three species are globally protected: the ferruginous duck, the white-tailed eagle, and the pygmy cormorant.

Plants: The swamp is home to 177 species of plants. During the 2004 study, a new plant species for the country, Elodea nuttallii (Nuttall's waterweed), was discovered in the area.

Fish: Sixteen species of fish have been identified in the swamp. The swamp's significant depth and stable water levels create ideal conditions for a rich fish fauna. Additionally, there is a shallow area within the swamp where cyprinid fish, such as carp, spawn.

Amphibians: The area supports two species of tailed amphibians and seven species of tailless amphibians, indicating a diverse amphibian population.

Reptiles: Ten species of reptiles have been documented in the protected area, adding to its rich biodiversity.

Mammals: The study identified 28 species of mammals within the swamp, highlighting its importance as a habitat for a variety of terrestrial animals.

This comprehensive biodiversity makes the Malak Preslavets Swamp a critical area for conservation efforts, serving as both a sanctuary for wildlife and a site of significant archaeological interest.

Fig.1

Fig.2

Fig. 3

2. Hydrobiological features of Srebarna lake

Lake Srebarna(fig. 4) is situated 18 kilometers from the town of Silistra, along the banks of the Danube River. The lake spans a water area of approximately 1,020 square kilometers. The lake's connection to the Danube River plays a crucial role in maintaining its ecological balance and biodiversity. Disruption of this connection has historically led to significant adverse effects on the lake's ecosystem.

In 1948, despite Lake Srebarna being declared a protected reserve, construction began on the Danube dike, stretching from Vetren to Silistra, with the aim of preventing the flooding of fertile lands (fig.5). This dike was completed in 1949, initiating a period of ecological decline for Lake Srebarna. The separation from the Danube caused a gradual degradation of the lake's environment, severely impacting its biodiversity.

Recognizing the detrimental effects of the dike, a section of approximately 500 meters in the northwestern part of the reserve was demolished in 1980, allowing the Danube's waters to once again enter the lake. This created an inter-dike space that partially restored the lake's connection to the river. However, this intervention was not sufficient to fully restore the lake's ecological health.

To further address the issue, a canal was constructed between 1993 and 1994 to reconnect the river and the lake. However, this canal only allowed water to flow one way—from the Danube into the lake during periods of high water levels. Recognizing the need for a more sustainable solution, a management plan for the reserve has proposed the construction of a new canal in the western part of the lake (fig.6). This canal would allow for two-way water exchange, enhancing the hydrological connection between the lake and the river.

In addition to these efforts, drainage systems have been installed to direct excess water from the lake back into the Danube. Unfortunately, these drainage systems have also contributed to the lake's current entanglement and the ongoing challenges in managing its water levels and overall ecological balance.

Fig. 4

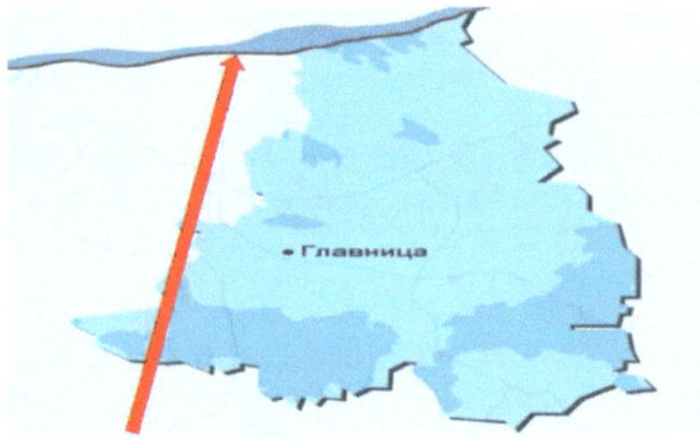

Fig. 5

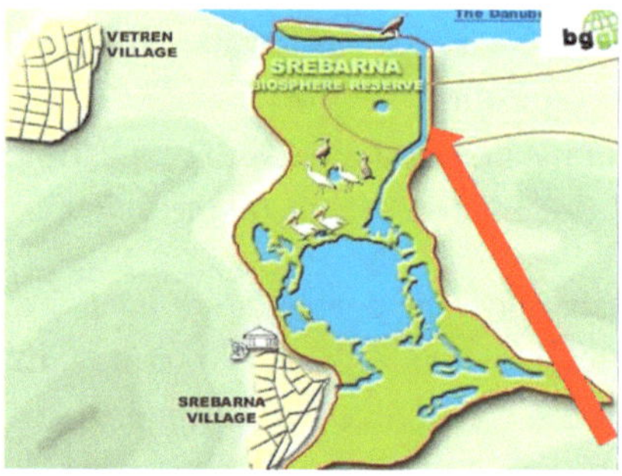

Fig. 6

3. Study of zooplankton in the two protected areas

3.1. Surveys in 2019

Visit to the Protected Area Malak Preslavets in April 2019

In early April 2019, we visited the Malak Preslavets Protected Area, along with a nearby reservoir that feeds its waters during periods of high water. During our visit, we took the opportunity to familiarize ourselves with the hydrological features of the swamp. We observed the dam constructed in 1997.

At the southwestern edge of the waterbody, a river flows into the swamp, originating from a series of small reservoirs in neighboring villages. To gain a better understanding, we also visited one such small reservoir near the nearby village of Zafirovo (fig.7).

Fig 7

Additionally, there are active underground karst springs in the area, as evidenced by the transparent patches observed in the winter ice, a phenomenon well known to local fishermen. During our discussions with these fishermen, they shared their observations on the fish populations in the swamp. They noted that the most abundant species are the common rudd, roach, silver carp, and perch. The swamp is also home to species like bitterling, weatherfish, and the spined loach. Interestingly, in this particular environment, the population of the common rudd is more dominant than that of the roach.

These observations, along with the input from the local fishermen, have provided valuable insights into the ecological dynamics of the Malak Preslavets swamp, highlighting the importance of both natural and man-made factors in sustaining its aquatic environment.

Sample Collection and Analysis

During our fieldwork, we conducted sample collections from both the Malak Preslavets swamp and a nearby micro-reservoir for chemical and biological analysis. For the zooplankton samples, we collected from the surface layer, specifically the top one meter, using a high-quality net with a mesh size of 90 microns (fig.8). The collected samples were preserved with formalin for later examination.

We also measured the water transparency in both the swamp and the reservoir using a Secchi disk (fig.9 and 10). This disk is a tool divided into alternating white and black sectors. When the disk is submerged into the water, the depth at which the colors of the disk become indistinguishable indicates the depth of light penetration. This depth is crucial as it reflects the extent to which sunlight can penetrate the water column, which is essential for photosynthesis and various biological processes within the aquatic ecosystem. A greater depth of visibility typically signifies clearer water with lower levels of suspended particles.

Fig 8

Fig. 9

Fig. 10

Our measurements revealed that the transparency layer in the swamp was 0.68 meters, while in the reservoir it was only 0.25 meters. This indicates that the swamp is undergoing a process of self-purification, whereas in the reservoir, this process has not yet commenced.

Following the sample collection, we performed a detailed analysis of the water samples based on several key indicators: ammonium ions, nitrates, nitrites, phosphates, and pH levels. These compounds, which include nitrogen and phosphorus, are essential biogenic elements and serve as critical indicators of the health of an aquatic ecosystem. The results of these analyses were compiled into a detailed table (table 1), providing a clear overview of the water quality in both the swamp and the reservoir.

The data from our analysis indicate the following: In the swamp, both nitrates and nitrites are present. The presence of nitrites, which are a byproduct of nitrate decomposition, suggests that the process of self-purification is actively occurring in the swamp. On the other hand, the absence of nitrites in the reservoir indicates that the self-purification process has not yet begun there. Additionally, there is no evidence of contamination with ammonium ions or phosphates in either the swamp or the reservoir, further confirming the relatively clean state of these water bodies.

Table 1

Indicator	Ammonium ions	Nitrates	Nitrites	Phosphates	pH
Malak Preslavets	0 mg/l	5 mg/l	0,10 mg/l	0 mg/l	8,3
reservoir	0 mg/l	10 mg/l	0 mg/l	0 mg/l	7,0

Zooplankton Sample Analysis and Observations

The zooplankton samples we collected were preserved with formalin. We conducted microscopic observations to analyze them.

Zooplankton are multicellular organisms that drift freely in aquatic environments. Despite having specialized organs for movement, they are carried along by water currents and cannot navigate against them, thus making them a passive component of the aquatic ecosystem. Their primary role is to aid in the process of water self-purification and to play a crucial part in the flow of energy within the ecosystem. Zooplankton consume biogenic substances, including compounds of nitrogen and phosphorus, which are vital for their growth and development. This feeding behavior positions them as a key element in the aquatic food chain, as they transform these nutrients into a form that can be utilized by higher trophic levels, such as fish and aquatic birds. By participating in these processes, zooplankton contribute significantly to the health and balance of aquatic ecosystems.

Phytoplankton serves as the primary food source for zooplankton, while zooplankton itself is a vital food source for newly hatched and adult fish, such as roach, common rudd, and bleak. These fish, in turn, are prey for birds, thus creating a complex and interdependent food web.

During our observations, we identified three major groups of zooplankton under the microscope, with the help of identification guides by Manuylova and Kutikova, as guided by Mrs. Tsavkova:

Rotatoria (Rotifers): Within this group, we observed species from the genus Asplanchna sp. (fig.11).

Cladocera (Water Fleas): We identified species from the genera Bosmina (fig12).

Copepoda: In this group, we observed the life stages from Nauplii, which hatch from eggs, to copepodites, and finally to mature adult copepods (fig.13).

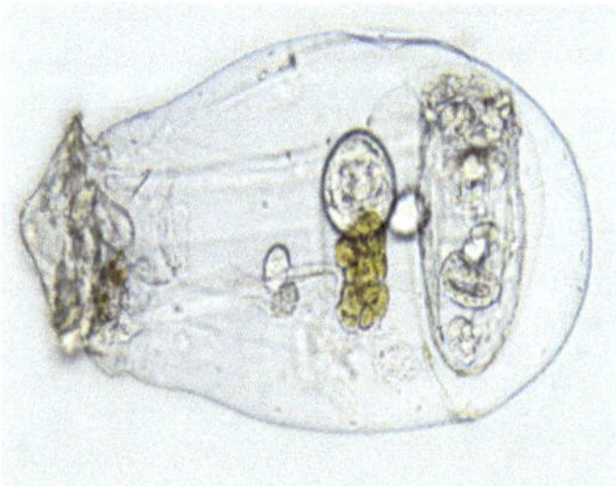

Fig 11

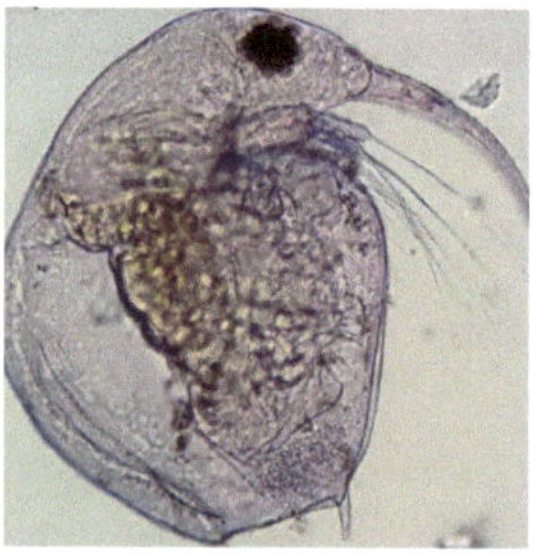

fig. 12

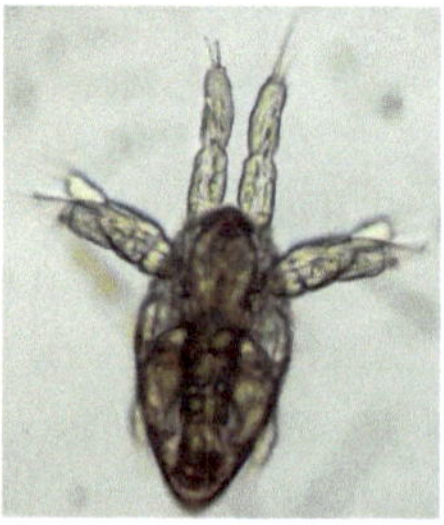

Fig. 13

This spring, the weather was unusually cold, with low temperatures, and as a result, the zooplankton we observed was characteristic of winter rather than spring. Winter zooplankton communities are significantly less diverse than those found in the warmer spring months. In comparison, a study conducted in 2004 identified a greater number of taxa in the swamp, likely due to warmer weather conditions that year.

Our observations highlighted the impact of seasonal variations on zooplankton diversity and abundance, and underscored the importance of zooplankton in maintaining the ecological balance of aquatic environments.

We determined that in the Malak Preslavets Swamp, there are so-called rotifer complexes characterized by the following ratio: 4:3:2 for the genera Brachionus, Keratella, and Polyarthra. According to Kutikova's classification, a rotifer complex consists of two or more genera with two or more species and/or subspecies. The observed ratios indicate the relative abundance and diversity of these rotifer genera within the swamp's zooplankton community(Diagram 1 and 2).

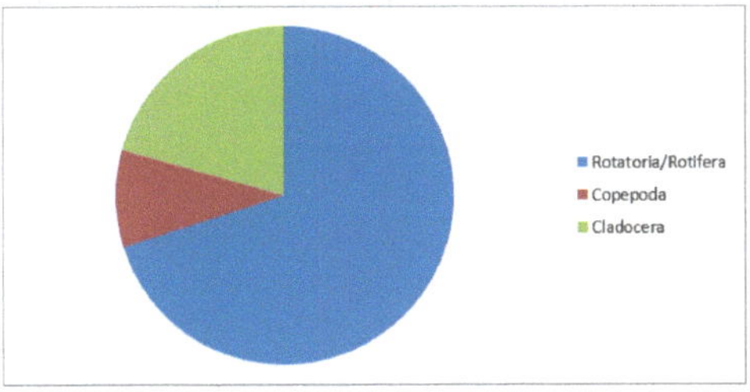

Rotatoria/Rotifera	68
Copepoda	9
Cladocera	20

Diagram 1

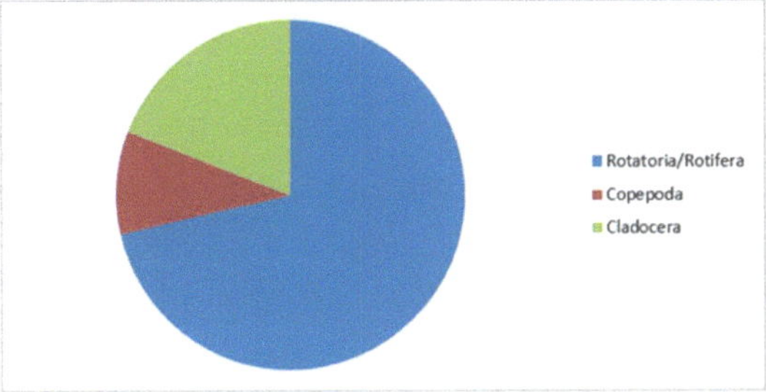

Rotatoria/Rotifera	15
Copepoda	2
Cladocera	4

Diagram 2

Conclusions

Water Quality and Self-Purification Processes:

Our analysis of the chemical and physicochemical indicators of the water reveals that self-purification processes are actively occurring in the Malak Preslavets Swamp. However, to gain a comprehensive understanding of the water quality, it is essential to examine different seasons and additional parameters. Unfortunately, there are contaminated areas within the swamp, with domestic waste and oil spills caused by irresponsible visitors, which impact the overall water quality.

Comparison with Srebarna and Hydrological Influences:

Unlike Lake Srebarna, the Malak Preslavets Swamp does not receive inflow from the Danube River due to the dam constructed in 1997. Instead, the swamp's water is sustained by surface runoff and karst springs, including a river that flows into the southern end of the swamp and karst springs. Both water bodies provide excellent feeding grounds for aquatic birds, indicating that the food chains in these ecosystems are beginning to resemble each other at the zooplankton level. Evidence of this similarity is found in the presence of rotifer complexes in both environments.

Zooplankton Observations and Insights:

3.1. Autochthonous Zooplankton:

The zooplankton in the swamp exhibits an autochthonous origin, meaning it self-reproduces within the swamp environment. This zooplankton provides a valuable food source for newly hatched fish, supporting the development of these fish populations.

3.2. Independence from the Danube:

The organization of rotifer communities in the swamp operates independently of the Danube River. This indicates that the swamp's zooplankton community is self-sustaining and not directly influenced by the Danube's ecological conditions.

3.3. Functional and Taxonomic Classification:

Kutikova's classification of rotifers not only reflects their taxonomic diversity but also their functional characteristics within the ecosystem. This classification provides insights into both the ecological roles and the taxonomic diversity of rotifers in the swamp.

These findings contribute to a deeper understanding of the ecological dynamics of the Malak Preslavets Swamp and highlight the importance of continued monitoring and conservation efforts to address pollution and preserve the delicate balance of these aquatic ecosystems.

Visit to the Protected Area Malak Preslavets in July 2019

On July 1, 2019, we revisited the Malak Preslavets Protected Area. During this visit, we collected additional samples for chemical analysis and conducted further microscopic observations. This follow-up allowed us to assess seasonal changes in water quality and zooplankton communities, providing a more comprehensive understanding of the dynamics within the protected area throughout different times of the year.

The results from the chemical analysis of the water can be seen in table 2.

Table 2

Indicator	Ammonium ions	Nitrates	Nitrites	Phosphates	pH
Malak Preslavets	0.1 mg/l	10 mg/l	0.050 mg/l	0.25 mg/l	8.1

The results from the hydrobiological analysis are represented in diagram 3.

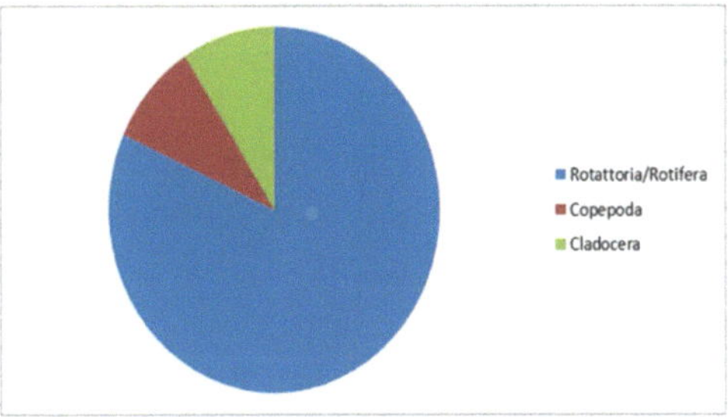

Rotattoria/Rotifera	36
Copepoda	4
Cladocera	4

Diagram 3

To ensure a more comprehensive study, we plan to continue our work on the project during the winter and spring seasons. This will allow us to conduct a thorough investigation and obtain more complete data on zooplankton species in both ecosystems.

The results of the analysis from Srebarna can be observed diagram 4.

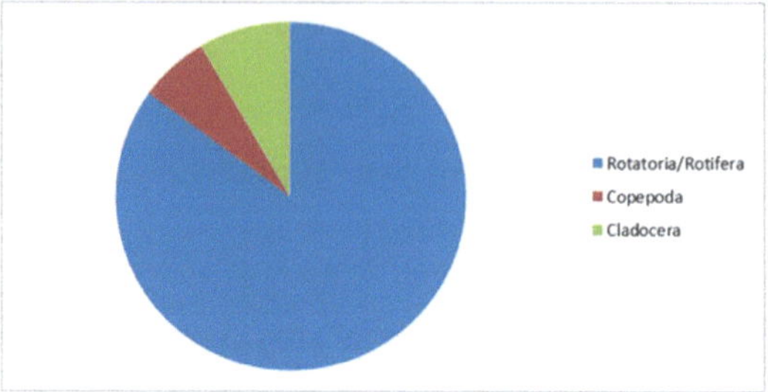

Rotatoria/Rotifera	40
Copepoda	3
Cladocera	4

Diagram 4

First three diagrams illustrate the species richness of zooplankton in the Malak Preslavets Swamp for the years 2004 and 2019, categorized by different zooplankton groups. The first two diagrams compare the diversity and abundance of these groups in the swamp during these two years, showing any changes or trends over time.

Diagram 1: Species richness of zooplankton groups in Malak Preslavets Swamp, 2004

Diagram 2: Species richness of zooplankton groups in Malak Preslavets Swamp, April 2019

Diagram 3: Species richness of zooplankton groups in Malak Preslavets Swamp, July 2019

Diagram 4: Species richness of summer zooplankton in Lake Srebarna, 2019

These diagrams provide a visual representation of the changes in zooplankton species richness over time in Malak Preslavets Swamp and offer a comparative perspective with the current state of zooplankton in Lake Srebarna.

The zooplankton species composition in Malak Preslavets Swamp and Lake Srebarna can be observed in Tables 3, 4, 5, 6, and 7.

	A	B	C	D
1	таксони (видове и подвидове)	MPr(2004)	MPr(july19)	Sr(july19)
2	ROTIFERA (ROTATORIA)			
3	Anuraeopsis fissa fissa (Gosse)		*	*
4	Asplanchna girodi Guerne	*	*	*
5	A. priodonta Gosse	*	*	
6	A. sieboldi (Leydig)			*
7	Brachionus angularis angularis Gosse	*		
8	B. a. bidens Plate	*	*	*
9	B. calyciflorus calyciflorus Pallas	*		
10	B. c. dorcas Gosse	*	*	*
11	B. c. spinosus Wierzejski	*	*	
12	B. diversicornis diversicornis (Daday)	*	*	*
13	B. d. homoceros (Wierzejski)	*	*	
14	B. falcatus Zacharias	*	*	
15	B. forficula forficula Wierzejski	*	*	*
16	B. f. minor Voronkov			*
17	B. f. voronkovi Fadeev			*
18	B. leydigii rotundus Rousselet	*		
19	B. quadridentatus ancylognathus Schmarda	*	*	
20	B. q. brevispinus Ehrenberg	*		
21	B. q. clunuorbicularis Skorikov	*		
22	B. q. rhenanus Lauterborn			*
23	B. q. quadridentatus Hermann	*		*
24	B. q. zernovi Voronkov	*		
25	B. urceus urceus (Linnaeus)	*	*	
26	Cephalodella catelina (Müller)	*		
27	C. hyalina Myers			*
28	Colurella oblonga Donner	*		
29	C. obtusa (Gosse)	*		

table 3

	A	B	C	D
30	C. uncinata bicuspidata (Ehrenberg)	*		
31	C. u. uncinata (Müller)	*		
32	Conochiloides dossuarius (Hudson)	*	*	
33	C. coenobasis Skorikov	*		
34	Encentrun (s. str.) marinum (Dujardin)	*		
35	Eosphora najas Ehrenberg	*		
36	Euchlanis dilatata dilatata Ehrenberg	*		
37	E. d. lucksiana Hauer		*	*
38	E. d. unisetata Leydig	*		*
39	Filinia longiseta longiseta (Ehrenberg)	*		
40	F. major (Colditz)	*	*	
41	F. passa (Müller)	*		
42	F. terminalis (Plate)	*	*	
43	Hexarthra mira (Hudson)		*	*
44	Keratella cochlearis cochlearis (Gosse)	*	*	*
45	K. c. hispida (Lauterborn)	*	*	*
46	K. c. tecta (Gosse)	*	*	*
47	K. quadrata quadrata (Müller)	*	*	*
48	Lecane (s.str.) balatonica Varga	*		
49	L. (s.str.) ludwigii (Eckstein)			*
50	L. (s. str.) luna luna Müller	*	*	*
51	L. (s. str.) l. presumpta Ahlstrom	*	*	*
52	L. (Monostyla) bula bula (Gosse)	*		*
53	L. (M.) bula diabolica (Hauer)	*	*	*
54	L. (M.) goniata (Harring et Myers)			*
55	L. (M.) lunaris (Ehrenberg)	*		*
56	L. (M.) perpusilla (Hauer)		*	*
57	L. (M.) quadridentata (Ehrenberg)	*	*	*
58	L. (M.) stenroosi (Meisner)	*		

table 4

	A	B	C	D
60	Lepadella (s. str) patella (Müller)	*		*
61	Lophoharis oxysternon (Gosse)	*		
62	Mytilina mucronata spinigera (Ehrenberg)	*		
63	M. ventralis redunca (Ehrenbsrg)	*		*
64	M. v. ventralis (Ehrenberg)	*		
65	Platyias polycanthus (Ehrenberg)			*
66	P. quadricornis quadricornis (Ehrenberg)	*		*
67	Polyarthra dolichoptera dolichoptera Idelson	*		*
68	P. euryptera Wierzejski	*		
69	P. longiremis Carlin		*	*
70	P. luminosa Kutikova	*		
71	P. mayor Burckhardt	*	*	*
72	P. vulgaris Carlin	*		
73	Synchaeta pectinata (Ehrenberg)	*	*	*
74	Testudinella patina intermedia (Anderson)	*	*	
75	T. p. patina (Hermann)	*	*	
76	T. p. trilobata (Anderson et Shepard)	*		*
77	T. semiparva Hauer	*		
78	Tetramastix opoliensis brevispina Ahlstrom	*		
79	T. o. opoliensis Zacharias	*	*	
80	Trichocerca (Diurella) antilopea (Pert)	*		
81	T. (D.) porcellus (Gosse)	*		
82	T. (D.) similis (Wierzejski)		*	*
83	T. (D.) taurocephala (Hauer)	*		
84	T. (D.) weberi (Jenings)	*		
85	Trichocerca (s. str.) capucina (Wierzejski et Zacharias)	*		*
86	T. (s. str.) elongata (Gosse)	*	*	
87	T. (s. str.) longiseta	*	*	
88	T. (s. str.) rattus carinata (Ehrnberg)		*	

table 5

	A	B	C	D
89	T. (s.str.) rattus rattus (Müller)	*		
90	Trichotria pocillum bergi (Meisner)	*		
91	T. p. pocillum (Müller)	*		
92	total	73	36	40
93				
94	COPEPODA			
95	Acanthocyclops capillatus (Sars)	*		
96	A. robustus (Sars)	*		
97	Eucyclops serrulatus (Fscher)	*		
98	Mesocyclops leuckarti (Claus)	*		
99	Thermocyclops crassus (Fischer)	*	*	
100	T. dybowskii (Lande)	*		
101	T. oithonoides (Sars)	*		
102	Micricyclops gracilis (Lilljeboborg)		*	*
103	nauplii	*	*	*
104	copepodites	*	*	*
105	total	9	4	3
106				
107	CLADOCERA			
108	Alona rectangula Sars	*	*	*
109	Bosmina longirostris (O. F. Müller)	*	*	*
110	Ceriodaphnia pulchella pseudohamata Bowkiewicz	*		
111	C. p. typica (O. F. Müller)	*		
112	C. quadrangula hamata Sars	*		
113	C. q. typica O. F. Müller	*		
114	Chydorus sphaericus (O. F. Müller)	*	*	*
115	Daphnia curvirostris Eylmann	*		
116	D. longiremis Sars	*		
117	D. longispina O. F. Müller	*		

table 6

118	D. pulex (De Geer)	*		
119	Daphnia sp. – ювенилна форма	*		
120	Diaphanosoma brahyurum (Lievin)	*		
121	Graptoleberis tetudinaria (Fischer)	*		
122	Moina micrura Hettich			*
123	M. rectirostris (Leydig)	*	*	
124	Pleuroxus trigonellus O. F. Müller	*		
125	P. uncinatus Baird	*		
126	Scapholeberis mucronata (O. F. Müller)	*		
127	Simocephalus vetulus (O. F. Müller)	*		
128	Syda crystallina (O. F. Müller)	*		
129	total	20	4	4
130				
131	общо за водоема	102	44	47

table 7

Conclusions:

Predominance of Rotatoria (Rotifera):

The majority of the zooplankton species richness in Malak Preslavets Swamp is composed of the group Rotatoria (Rotifera). This indicates that rotifers are the dominant zooplankton group in this environment.

Comparison of Spring and Winter Zooplankton:

The second diagram illustrates the species richness observed in April, a spring month. However, in 2019, April was unusually cold, leading us to capture winter zooplankton instead. We compared the species richness from 2019 with that from 2004 to analyze any changes or differences.

Summer Zooplankton:

The third diagram represents the zooplankton composition from July, which reflects the summer season. During this time, warm-loving rotifers are predominant, indicating seasonal variations in the zooplankton community.

Comparison Between Lakes:

Diagrams 3 and 4 are notably similar, showing comparable species richness between the two lakes. This similarity is surprising given that Lake Srebarna and Malak Preslavets Swamp have different water supply sources—Malak Preslavets receives surface (non-Danube) and groundwater, while Lake Srebarna is fed by the Danube and groundwater. This raises questions about the origin of zooplankton communities: autochthonous (self-reproducing in Malak Preslavets) versus allochthonous (originating from the Danube in Srebarna), with Srebarna exhibiting a mixed community.

Surveys in 2020

On July 31, 2020, we undertook a field visit to both protected areas—Srebarna Lake and Malak Preslavets Swamp.

Following the fieldwork, we conducted detailed analyses of the collected samples. This involved creating diagrams to represent the number and biomass of zooplankton in both Srebarna Lake and Malak Preslavets Swamp. To accurately capture the species diversity and quantify the zooplankton, we performed microscopic observations. We used a specialized counting vessel known as a "camera," which features a grid where a portion of the water sample is placed. This device allows for precise counting of zooplankton species by facilitating a systematic examination of the sample.

Statistical methods (R- statistics) were employed to process the numerical data obtained from these observations, enabling us to produce comprehensive diagrams that illustrate the distribution and abundance of different zooplankton groups.

The results for Srebarna are in diagrams 5 and 6.

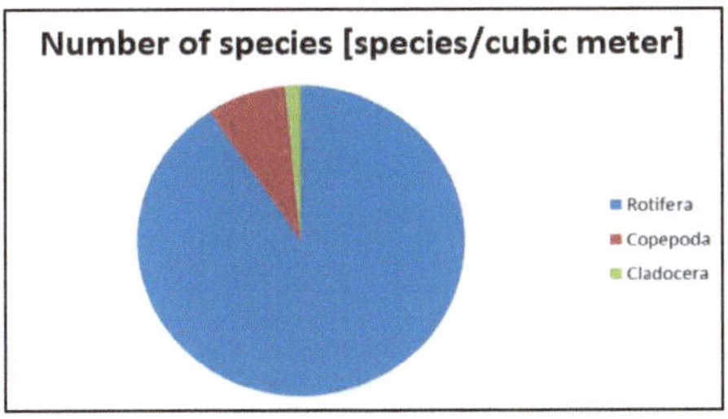

Diagram 5

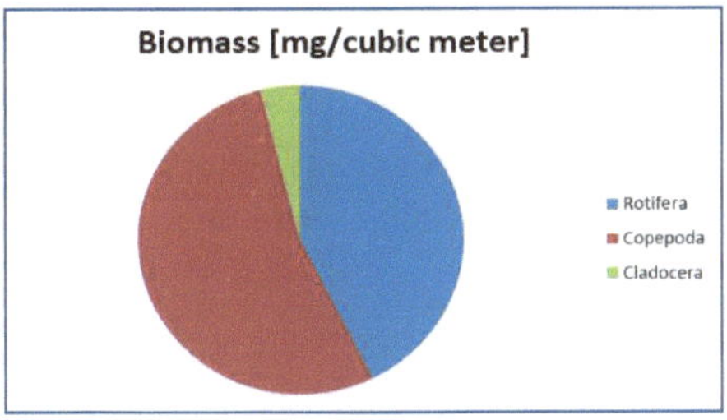

diagram 6

The results for Malak Preslavets are in diagrams 7 and 8.

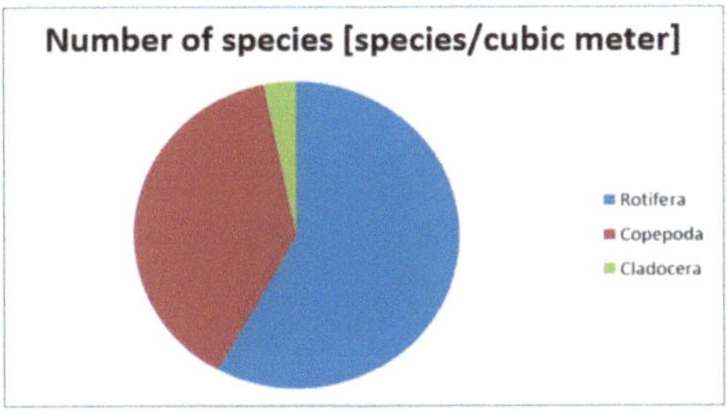

Diagram 7

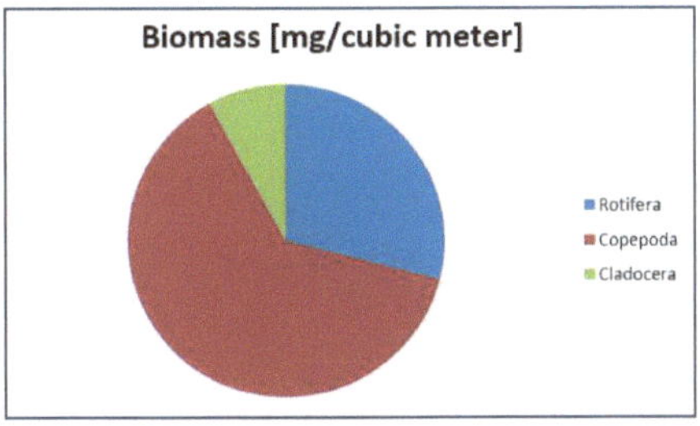

Diagram 8

The resulting diagrams reveal some notable trends. In Srebarna Lake, the Rotifera group is found to be the most numerous among the zooplankton, while the Copepod group exhibits the highest biomass. Interestingly, a similar pattern is observed in Malak Preslavets Swamp, where Rotifera also dominate in terms of species count, and Copepods have a significant biomass. These findings suggest that despite differences in the hydrological characteristics and water sources of the two protected areas, the zooplankton communities share comparable structural traits. This indicates that similar ecological processes may be influencing zooplankton distribution and abundance in both ecosystems.

Surveys in 2021

To facilitate a more comprehensive comparison of species diversity between the two protected areas, we continued our research into 2021. This year, we aimed to gather additional data from the central part of Lake Srebarna, which required obtaining a special permit from the Ministry of Environment and Water due to the lake's status as a protected reserve. Although there was a slight delay, the permit was eventually granted.

In early May 2021, we revisited Lake Srebarna. With assistance from the reserve guards, we accessed the lake by boat. Under the expert guidance of Mrs. Veselka Tsavkova, we collected samples for hydrobiological analysis from two specific locations: the pier and the central part of the lake. During this visit, we also conducted transparency measurements using a Secchi disk (fig.14 and 15). Our findings revealed that the water transparency was quite substantial, indicating a period of "pure water." The transparency depth was measured at 1.90 meters in the central part of the lake and 1.30 meters at the pier.

Fig. 14

Fig.15

In addition to our work at Lake Srebarna, we also conducted observations at Malak Preslavets Swamp. During this visit, we examined the dike structure and collected two types of samples for hydrobiological analysis: one set for live zooplankton and another fixed with formalin. We again used the Secchi disk to measure water transparency, which revealed that sunlight penetrates to a depth of 0.95 meters in the swamp.

Following sample collection, we analyzed the data in collaboration with Mrs. Tsavkova. One of the most intriguing aspects of our analysis was the microscopic observation of live zooplankton, which we documented on video for further study. We performed additional microscopic observations and species counting using a specialized counting vessel known as a Bürker chamber (fig.16). The numerical data obtained were processed using statistical methods to generate comprehensive diagrams illustrating the findings.

Fig. 16

The diagrams presented below offer detailed insights into the zooplankton communities of both protected areas. Diagrams 9 and 10 represent the data for Lake Srebarna, while diagrams 11 and 12 depict the findings from Malak Preslavets Swamp. These visualizations will aid in the comparison of species diversity and biomass between the two ecosystems, contributing to a better understanding of their respective aquatic environments.

The diagrams show that in Srebarna the Cladocera group has the largest share in the number and biomass. In Malak Preslavets the representatives of Copepod group have the largest share in number but the largest share in biomass have the representatives of Cladocera group.

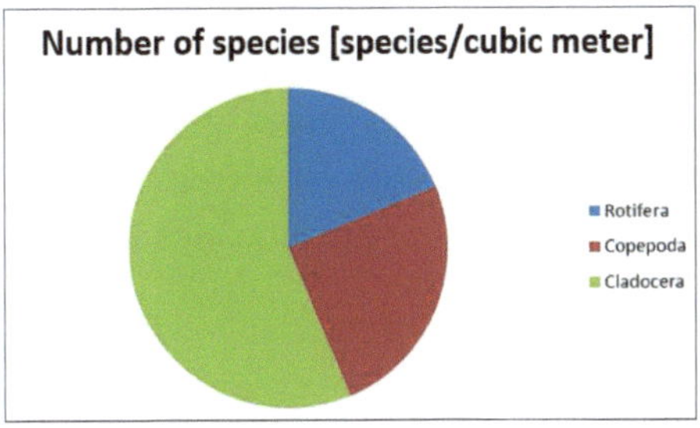

diagram 9

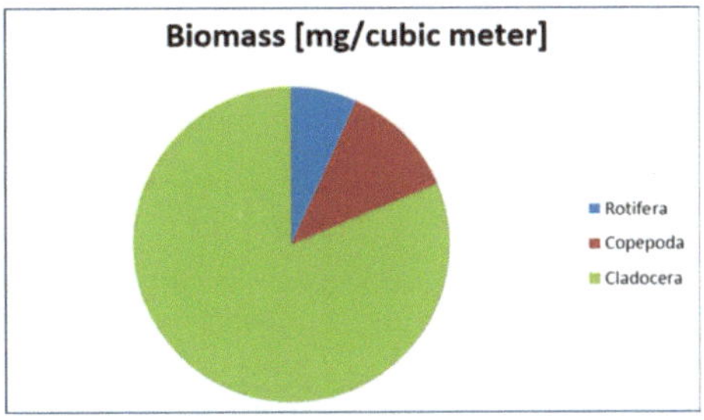

diagram 10

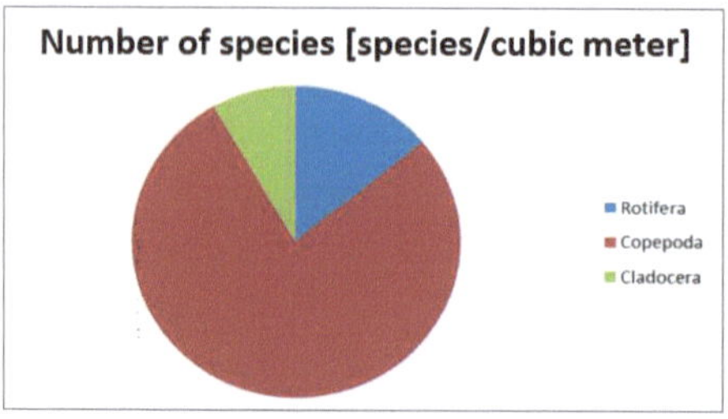

diagram 11

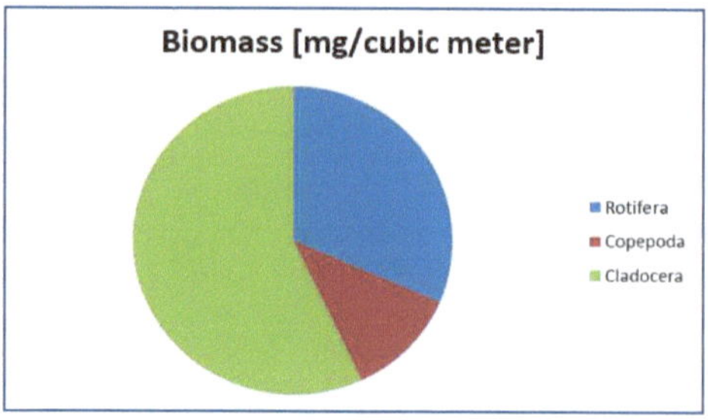

diagram12

Conclusions:

1. The predominant part of the species richness of Lake Malak Preslavets and Srebarna is composed of the Rotatoria Group (Rotifera).

2. The reasons for the values obtained can be established by statistical methods and in the case of very frequent sampling. The reasons are multifactorial - temperature, chemical composition of water, sunshine, wind, relation with micro-dams, etc.

3. The diagrams for Malak Preslavets and Srebarna are very similar - they reflect the species richness of the two lakes. This is very surprising, because Srebarna and Malak Preslavets have different ways of water supply in relation to the Danube River - Malak Preslavets from surface water (not from the Danube) and groundwater, and Srebarna - from the Danube and groundwater. The issue of the emergence of zooplankton communities is discussed. In Malak Preslavets they are autochthonous (self-generating in the swamp). In Lake Srebarna they are mixed, i.e. there is autochthonous in the lake itself and allochthonous (brought from the Danube).

3.4 Surveys in August 2021

In August, we resumed our research and visited both protected areas once again. To collect samples for further analysis, we entered the lake by boat and focused on sampling from the central regions of both reservoirs. During this fieldwork, Mrs. Veselka Tsavkova recommended adopting a novel scientific approach for analyzing our samples. She introduced us to the work of Torvald Sørensen, a Danish botanist renowned for his contributions to ecological research.

In his seminal 1948 study, Sørensen developed the coefficient of species similarity as a tool for comparing the species compositions of different aquatic environments. This coefficient provides a quantitative measure of how similar the species assemblages are between various reservoirs or ecosystems. By applying Sørensen's coefficient, we aimed to gain deeper insights into the similarities and differences in species diversity between Lake Srebarna and Malak Preslavets Swamp.

Sørensen's approach focuses on comparing the presence and abundance of species across different locations, allowing researchers to assess the degree of ecological similarity or divergence. This method has become a standard tool in ecological studies, helping scientists understand how various factors influence biodiversity and community structure over time and across different environments.

By incorporating this method into our analysis, we sought to enhance our understanding of the ecological relationships between the two protected areas and to provide a more robust comparison of their zooplankton communities.

The coefficient is marked with "K" as:

$$K = 2c/(a+b), \text{ where:}$$

- "c" is the number of matching species for both reservoirs;

- "a" is the number of species in one reservoir;

- "b" is the number of species in the other compared reservoir.

The value of the coefficient varies between 0 and 1. Coefficient 0 means that there is no similarity.

Mrs. Tsavkova provided us with her data from her researches from 2004 and 2006, where the coefficient of species similarity for Lake Srebarna and the Malak Preslavets swamp was calculated. We decided to compare them with the results we obtained in 2021.

Diagram 13 shows the coefficient of species similarity for 2004 for the spring-summer season. It can be seen that the coefficient changes, although slightly, over two seasons.

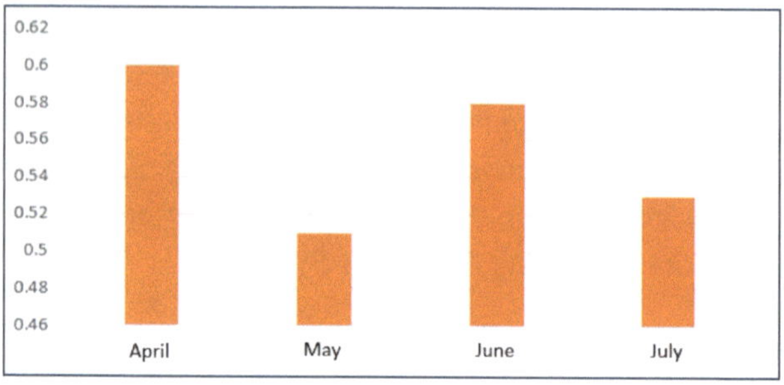

Diagram 13

Based on the results obtained, we can calculate the coefficient of species similarity for 2006 and 2021:

$$K(2006)= 2c/(a+b)= 2x31/(81+51)= 0.46$$

$$K(2021)= 2c/(a+b)= 2x14/(32+21)= 0.53$$

Diagram 14 represents the data of the comparison of the coefficient in 2006 and 2021.

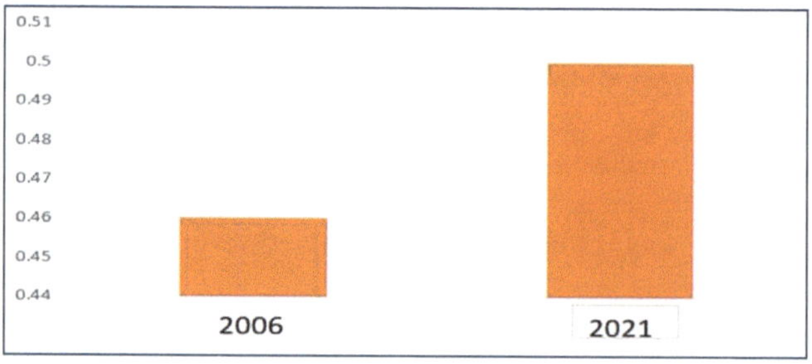

Diagram 14

Comparison of the zooplankton from 2006 and 2021

Table 8 contains data on the species by groups for the lake and the swamp for 2006 and 2021. It shows the species, their number in Srebarna Lake, the number of species in Malak Preslavets swamp and the number of matching species in both reservoirs.

Table 8

	Year	2006			2021		
		Srebarna	*M. Preslavets*	*Matching species*	*Srebarna*	*M. Preslavets*	*Matching species*
Group	*Cladocera*	*14 species*	*9 species*	*5 species*	*5 species*	*1 specie*	*1 specie*
	Copepoda	*11 species*	*6 species*	*5 species*	*3 species*	*2 species*	*2 species*
	Rotatoria	*56 species*	*36 species*	*21 species*	*24 species*	*18 species*	*11 species*
Total		***81 species***	***51 species***	***31 species***	***32 species***	***21 species***	***14 species***

Conclusions:

Increase in Sorensen Coefficient: Over the past 15 years, there has been an increase in the Sorensen coefficient, which indicates a growing similarity in species composition between the two protected areas, Lake Srebarna and Malak Preslavets Swamp.

Comparison with Historical Data: The values of the Sorensen coefficient for the years 2006 and 2021 show a level of similarity that is comparable to the species diversity observed during two distinct seasons in 2004. This suggests that the biodiversity of the two reservoirs has remained relatively consistent over time, despite fluctuations in seasonal conditions.

Minimal Difference in Coefficient Values: The small variation in the Sorensen coefficient values further confirms that the two reservoirs are quite similar in terms of species diversity. This minor difference highlights that the overall species composition in both environments has remained stable over the years.

Reasons for Species Similarity:

4.1 Stable Water Supply: There have been no changes in the methods of water supply to either reservoir. Both Lake Srebarna and Malak Preslavets Swamp continue to receive water through their established natural and artificial sources, which supports the maintenance of similar aquatic communities.

4.2 Consistent Physical and Chemical Parameters: The physical and chemical characteristics of the water in both reservoirs have remained stable, with no evidence of industrial or domestic pollution. This stability in water quality contributes to the preservation of similar ecological conditions in both environments.

4.3 Absence of Natural Disasters: No significant natural disasters have affected either reservoir during the period under study. The lack of such disturbances has allowed the ecosystems to develop and maintain similar species compositions over time.

Sources used:

1. Chislenko, L. L. (1968): Nomogms for calculation of the body mass of water organisms through body size and form.-106 pp., (Nauka Publishing House) Moskow;

2. Damian-Georgesku, A. (1966); Fauna Republicii Socialiste Romania, yol.IV, fasc, 8, Calanoida.-130 Psg... (Intreprinderea Pligrafica Infomatia), Bucuresti;

3. Kutikova. L. A. (1970): Rotifers from the fauna of USSR (Rotatoria). Subclass Eurotatoria. Orders: Ploimida, Monimotrochida, Pacdotrochida.-744 pp., (Nauka Publishing House) Leningrad;

4. Manuylova, E. F. (1964): Branchiopod crustaceans (Cladocera) from the fauna of the USSR.-327 pp., Nauka Publishing House) Moskoy. Leningrad;

5. Monchenko, V. I. (2003): Free-living cyclopoid copepods of the Ponto-Caspian basin.-350 pp., (Naukova Dumka House) Kiev;

6. Tsavkova, V. (2005): Comparative analysis of the spring-summer zooplankton communities of the Srebarna and Malak Preslavets lakes. - Scientific papers, volume 44, series 7,1, May scientific readings of RU "Angel Kanchev", ISSN-1311-3321, page 172-183.

www.ingramcontent.com/pod-product-compliance
Lightning Source LLC
LaVergne TN
LVHW051131210926
848482LV00001B/53
* 9 7 9 8 8 9 2 4 8 5 5 0 0 *